IMAGES
of Aviation

Beaumont's Civil Air Patrol in World War II

Young and old, black and white—everyone worked together to make Base 10 a success. Pictured standing at center is 73-year old William F. Van Cleave, who leads a group in stretching fabric over the exterior of an airplane. Beaumont boasted Civil Air Patrol (CAP) members who ranged in age from 14 to 73. (Courtesy of the Larry Jene Fisher Collection, Lamar University.)

ON THE COVER: An unidentified CAP member is "hand-propping" to start an aircraft. Many aircraft of that era had either a limited or nonexistent electrical system. Without a starter, hand propping was necessary to get aircraft started. Anyone attempting it needs to be properly trained; while it can be performed safely, an accident can lead to severe injury, including loss of limb, and, in some cases, death. (Courtesy of the Larry Jene Fisher Collection, Lamar University.)

Images of Aviation
Beaumont's Civil Air Patrol in World War II

Penny L. Clark

ARCADIA
PUBLISHING

Copyright © 2021 by Penny L. Clark
ISBN 978-1-4671-0620-7

Published by Arcadia Publishing
Charleston, South Carolina

Printed in the United States of America

Library of Congress Control Number: 2021934604

For all general information, please contact Arcadia Publishing:
Telephone 843-853-2070
Fax 843-853-0044
E-mail sales@arcadiapublishing.com
For customer service and orders:
Toll-Free 1-888-313-2665

Visit us on the Internet at www.arcadiapublishing.com

Thanks to my brother Alan Clark and his wife, Sharon, for their love and support and the memory of my parents, Fred and Anemone Clark, and my beloved sister Elizabeth Clark Moyer.

Contents

Acknowledgments		6
Introduction		7
1.	Wings over the Gulf	11
2.	Spit and Polish	39
3.	Tragic Losses and New Solutions	55
4.	Partners for Victory	69
5.	A Nation at War	85
6.	The Renaissance Man of East Texas	109
7.	No Dull Boys at Beaumont	117
8.	Moving on to New Work	123
Bibliography		126
About the Organization		127

Acknowledgments

I would like to express my gratitude to all the people and organizations that assisted me in this project. Most significant is Lamar University, which allowed me to use the Civil Air Patrol photographs from the Larry Jene Fisher Collection; all but three of the photographs in the book are from that collection. The university also gave me the opportunity to write much of the book during work hours. Thanks go to my colleagues in Lamar's Special Collections, Charlotte Holliman and David Worsham, for scanning photographs, brainstorming ideas, and listening to endless stories about the Civil Air Patrol. Judith Linsley, director of the Center for History and Culture of Southeast Texas and the Upper Gulf Coast at Lamar University, whose father John H. Walker served with the Civil Air Patrol, proofread the manuscript.

A big thank-you goes to the Martin and Osa Johnson Safari Museum in Chanute, Kansas, which allowed me to use two photographs of the Johnsons with the *Spirit of Africa and Borneo* in this book.

Larry Kelley—a commercial pilot, aircraft mechanic, and now a local maritime manager—shared his encyclopedic knowledge of antique aircraft as well as local landmarks. Historian Greg Riley identified the fabulous automobiles.

Special recognition goes to Frank A. Blazich Jr., PhD, curator of modern military history for the Division of Political and Military History at Armed Forces History at the Smithsonian Institution's National Museum of American History and director of the Colonel Louisa S. Morse Center for CAP History and national historian emeritus, was an invaluable resource. Frank provided information from the Morse Center, suggested avenues for further research, and proofread the manuscript.

A unique thank-you goes to my brother Alan Clark, who nudged me to complete the book and even wrote a fictional account of the famed aircraft *Spirit of Africa and Borneo*.

Introduction

Although Base 10 of the Civil Air Patrol, located near Beaumont, Texas, lasted only a short time, from June 1942 to August 1943, its story reveals much about Texas and the nation during the war.

Beaumont and all Southeast Texas played a key role in victory for the Allies. It was vital to have massive quantities of petroleum to fight a war in the 1940s. As Soviet leader Joseph Stalin put it, "This is a war of engines and octanes."

Beaumont had unleashed the petroleum age in 1901 when the famed well at Spindletop south of the city exploded with an incredible 100,000 barrels of oil per day. Previously, a well that produced 50 barrels a day was noteworthy. Moreover, Beaumont's location near the Gulf of Mexico enabled easy transportation of "black gold" to the rest of the world. Although Spindletop was soon eclipsed by other boomtowns as an oil-drilling capital, Beaumont and Southeast Texas remained vital as an oil refining center.

During World War II, the US Army Air Forces on average would use as much gasoline in a day as was consumed during the entire first world war. The Germans knew about the importance of oil coming from Texas and sent submarines to sink vessels containing products that would aid the Allied war effort, particularly vulnerable oil tankers.

The US Navy was spread way too thin along the 3,700-mile coast of the Atlantic and Gulf of Mexico to keep the German subs at bay. In 1942, U-boats sank 12 vessels in January and 42 in March. By May, the losses were so bad that the government released no statistics.

With the destruction of the ships, there was not only loss of precious materials but also human life. The US Merchant Marine, the civilian sailors who crewed the ships carrying vital goods, lost nearly four percent of its members during the war. Significantly, the odds of dying in the Merchant Marines were greater than in the actual uniformed services. Southeast Texas lost 146 Merchant Marines: brothers, sons, and husbands serving on ships that were destroyed.

How was the United States going to combat the menace of the fearsome submarines? One way was creating the Civil Air Patrol to utilize civilian airplanes and pilots. But many military leaders questioned the ability of civilians who owned small airplanes to make a difference.

Gill Robb Wilson, a founder and a proponent of the CAP, stated, "We may not sink any submarines, but we may be able to frighten them into staying below the surface. That would reduce their spread and the accuracy of their fire would suffer, giving our shipping a fighting chance. They wouldn't be knocked over like sitting ducks as they are now."

The CAP had two staunch allies—oilmen and Merchant Marine unions—who recognized that it would be beneficial to their interests. Two Texans—George Haddaway, secretary of the Texas Private Flyers Association and publisher of *Southern Flight*, an aviation magazine, and oilman David Harold "Dry Hole" Byrd—were major advocates of the patrol. Byrd would become the wing commander of the Texas Wing of CAP, with Haddaway serving in different capacities, including executive officer and public relations director. The Civil Air Patrol was founded only six days before Pearl Harbor, on December 1, 1941, and legally established on December 8, 1941.

The first two bases, in New Jersey and Delaware, were opened on a 30-day trial basis, with the time extended as the bases proved successful. Over time, CAP boasted 21 bases from Maine to Mexico, with three in Texas at Beaumont and Brownsville, and later, San Benito and Corpus Christi.

The Civil Air Patrol was indeed a unique organization as it was semi-military in nature. It was composed of civilians who were often too young or too old or had disabilities disqualifying them for regular military service. Most of the men and women had experience flying in civilian life, and many brought their personal aircraft, which they flew in the CAP. Other men had other useful skills, including medical care, navigation, airplane mechanics, radio operation and repair, and photography.

Organizing for the Civil Air Patrol in Beaumont began in January 1942. H.A. Smalley, who managed the Beaumont Municipal Airport, was named group commander in January 1942. Smalley worked to sign up civilian pilots for the CAP. After his tragic death on March 17, 1942, in an airplane crash, there were other men involved with the group, and William M. "Bill" Cason was briefly base commander. He was removed when CAP members of the base signed a petition asking for his removal. He was replaced by George Haddaway, a 33-year-old who had a lifelong interest in aviation and good looks and was public relations savvy.

Coastal Patrol Base No. 10 of the Civil Air Patrol, located at Beaumont's Municipal Airport, then 10 miles west of the city, at Amelia, a suburb of Beaumont, was officially activated on June 24, 1942, but had already flown missions at the request of ship captains. Before serving in the Civil Air Patrol, the members had to undergo 80 hours of training. They had to hone their skills in flying and navigation as well as learn new tasks, such as marching and military protocol. The first official patrol from the base was on July 7, 1942, with William M. "Bill" Cason as the pilot.

The boundaries of Base 10's patrols were halfway between Grand Isle, Louisiana (Base No. 9) and Galveston, and Corpus Christi (Base No. 15) and Galveston.

Haddaway had a difficult task, as he was sending men to do a dangerous job. Every mission the men went on could be their last. They were flying single-engine aircraft over water at very low altitudes so they could locate submarines or survivors of a sub attack. They also flew over convoys of ships carrying supplies vital to the Allies. The tiny aircraft went out in pairs, calling the other aircraft "sister ship." They were equipped with two-way radios and kept in frequent communication with the base, reporting their location once every 30 minutes. For the most part, there was little adventure. Men spent hours and hours of routine flying, covering ships bringing vital goods for the war.

But every time they climbed into those tiny airplanes, they were risking their lives. Four men were killed while serving at Base 10, and five others joined the Duck Club, or men who ended up in the water due to mechanical failure. The base also lost an amphibious aircraft, a Sikorsky-39, the famed *Spirit of Africa and Borneo*, which was used by celebrated explorers, filmmakers, and authors Osa and Martin Johnson on their travels through Africa and Asia. The plane, which could float in the water or fly through the skies, was at Base 10 less than a month before it was lost in the Gulf.

The Civil Air Patrol was a quintessentially American outfit. While a Beaumont newspaper described it as a "smoothly running organization clicking off its duties with clock-like precision," that was not the whole story. It was disorganized, especially in the beginning. There was even an argument over whether it was part of the Army or Navy. The pay was modest—only $8 a day for pilots and $5 a day for other employees such as guards and secretaries. But even worse, even these small paychecks were late in arriving.

As Base 10 commander George Haddaway recollected many years later, "We couldn't fight our way out of a wet paper sack. I couldn't get guns to train my men in the local squadron. We used wooden guns, we had a hell of a time with no pay, and most people had to live on their eight dollars a day allowance."

While there were challenges, people with a wide array of talents worked together to make the base a success. One of the most significant was Charles Didio, a hairdresser extraordinaire in civilian life, who took a ragtag gang of men older and less fit than traditional servicemen and whipped them into a professional corps. The base boasted members both young and old—William Van Cleave, who was an "A&E" (aircraft and engine) mechanic, and Jimmy Marshall, a 14-year old who served as a messenger and provided inspiration for the entire base.

Base 10 did attract a wide array of Americans. An interesting group was four who came from Emporia, Kansas, who loaded up a 1940 Chevy coupe and came to Beaumont to make extra money and help the war effort. They included a radio shop operator, a 16-year-old high school student, and two college professors who gave up their classrooms, where they lectured on journalism and Dante's effect on Chaucer.

While some at Base 10 were from out of state, many were native Texans from Beaumont, Dallas, or Wichita Falls. They included ranchers, oilmen, bankers, and even a mortician. They ranged from rich to poor. George Haddaway remembered, "I found out that the best pilots and the best men I had were not millionaires. In fact, the millionaires didn't stay long. The best men I had were guys who came up the hard way and had worked all their lives."

And it was not only white men who kept the base going. Despite the horrors of the Beaumont Race Riot in the summer of 1943, African Americans worked at the base, where they served as mechanics, cooks, and janitors. Although women were banned from coastal patrol duty, some women in the CAP did fly missions delivering vital papers, ferrying airplanes, and chauffeuring VIPs. Others learned valuable skills that enabled them to serve in the WASPs, or Women Airforce Service Pilots, later in the war, which freed men for combat.

There were only a handful of women at Base 10, and none were pilots, observers, mechanics, or intelligence officers. Instead, they lent their skills to accounting, maintaining the plotting board (a display of the location of airplanes on patrol), and doing the mountains of paperwork necessary for maintaining a government agency. Some women in CAP served as radio operators and manned base control towers.

As the war progressed, the submarine menace lessened and the need for antisubmarine flights ended. The last flight from Base 10 was on August 31, 1943. Pilots from Base 10 had flown a total of 14,000 hours over the water at great risk.

Perhaps the most rewarding moments for Base 10 pilots was when the sailors on a ship would tear off their shirts and wave them in salute of the CAP who had not only saved the lives of countless Merchant Marine sailors but also allowed precious Texas oil to reach the Allies and literally fuel the way to victory.

The Civil Air Patrol, and its talented members, did not end their service in August 1943, when tiny aircraft were no longer on submarine patrol. Some members of Base 10 continued in the military, where they often utilized their aviation skills. In Texas, the CAP fought forest fires, patrolled the Texas-Mexico border, and trained youngsters to fly. And the good work did not end after the war. Today, the group conducts search and rescue, transportation of critical personnel and supplies, and educating young people.

Unfortunately, the members of the Civil Air Patrol did not receive benefits after the war. They were considered volunteers and therefore ineligible for programs other veterans enjoyed.

In 1949, the work of the CAP was recognized by an Air Medal presented on behalf of Pres. Harry S. Truman. The medal was awarded by the US Air Force to coastal patrol personnel with at least 200 hours of coastal patrol flight time. The citation read,

> For meritorious achievement while participating in anti-submarine patrol missions during World War II. The accomplishment of these missions in light commercial type aircraft despite the hazards of unfavorable weather conditions reflects the highest credit upon this valiant member of the Civil Air Patrol.
>
> The high degree of competence and exceptional courage he displayed in the volunteer performance of a hazardous and difficult task contributed in large measure to the security of coastal shipping and military supply lines.
>
> His patriotic efforts aided materially in the accomplishment of a vital mission of the army air force in the prosecution of the war.

In 2014, approximately 200,000 World War II Civil Air Patrol members were awarded the Congressional Gold Medal for flying armed and humanitarian missions. Now both men's and women's contributions were recognized, with an illustration showing a duo outfitted for a coastal patrol mission, with the woman wearing a flying suit, a soft flying helmet, and goggles, while her colleague is wearing a garrison cap and holding binoculars.

But perhaps the greatest award was not presented by the government but from their own insight into the work they did. As George Haddaway recollected, "But as tough as things were, and as bad as our airplanes and our engines and our sorry communications were, we still had a tremendous effect on submarines along our coastal waters. . . . So nobody is ever going to take that away from us."

One

WINGS OVER THE GULF

George E. Haddaway, the 33-year-old commander of the base, recollected, "My main function at the base was keeping things going. I checked on patrol reporting procedures, the condition of our equipment, and handled everyone's complaints and suggestions. I knew my men were accomplished pilots and observers, but every now and then I would ride in the back seat to let them know that I was taking some risks too."

Base 10 of the Civil Air Patrol was 10 miles west of town off Highway 90. The patrol made use of the Beaumont Municipal Airport, with construction beginning in 1932 and activation on October 1, 1937. It consisted of a terminal building and three shell-surfaced runways, 100 feet wide and between 2,500 and 3,500 feet long. The first floor of the terminal housed the airport office, a canteen, and a Civil Aeronautics Administration (CAA) office. CAP offices were on the second floor. A large all-metal hangar and a former Civilian Conservation Corps dormitory used as a recreation building were nearby. After the war, it provided commercial service for Trans-Texas Airways and Delta Airlines until 1946, when the Jefferson County Airport (today's Jack Brooks Airport) opened. The convertibles below are a 1940 Buick Special on the left, owned by Lt. Charles Kehoe, and a 1941 Ford, owned by a Lieutenant Johnson.

William Mabry "Bill" Cason enjoyed a full life in the aviation field. In 1942, he briefly served as the base commander, but was ousted from that position when the men under his command signed a petition asking for a new base commander. While it is unknown what the issue was, it was not a lack of flying experience. Cason had been a pilot since 1919 and had worked as the private pilot for Al Buchanan, a drilling contractor in San Antonio. He is pictured here with a Beechcraft Staggerwing and his wife, Lucille, who could herself boast of 2,000 hours flying time. Cason later served as commanding officer of the CAP in San Antonio, recruiting young men for the Army Air Corps. By the 1950s, he had amassed 15,000 flying hours and worked as a salesman for Alamo Aviation, selling $400,000 worth of airplanes in a year.

DEPARTMENT OF THE AIR FORCE

CERTIFICATE OF HONORABLE SERVICE

BE IT KNOWN THAT

Lawrence O. Fisher, 8-1-2001

SERVED WITH THE ARMED FORCES OF THE UNITED STATES DURING WORLD WAR II
AS AN ACTIVE DUTY MEMBER OF THE CIVIL AIR PATROL
(A VOLUNTEER CIVILIAN AUXILIARY OF THE ARMY AIR FORCES)
AS A

BELLIGERENT

AS DEFINED IN ANNEX TO HAGUE CONVENTION No. IV, OF OCTOBER 18, 1907

WASHINGTON, D.C.
15 MAY 1948

FOR THE SECRETARY OF THE AIR FORCE

MAJOR GENERAL, UNITED STATES AIR FORCE
NATIONAL COMMANDER, CIVIL AIR PATROL

Hoyt S. Vandenberg
CHIEF OF STAFF, UNITED STATES AIR FORCE

Lawrence "Larry" Fisher's certificate of honorable service as a belligerent is seen here. If CAP personnel had not been considered belligerents, when captured by the enemy they would have been considered guerillas and possibly killed. As belligerents, they would have been treated as prisoners of war. The Civil Air Patrol was neither fish nor fowl or, as the local commander George E. Haddaway termed it, "Neither swans nor gooses, but swooses."

Joe Wier is prepared for a flight with his binoculars and life vest, which was dubbed a "Mae West," in honor of the popular, well-endowed actress. CAP coastal patrol flights frequently flew low, at only several hundred feet altitude, protecting convoys of ships carrying desperately needed supplies, including petroleum, to other parts of the United States and its Allies in Europe.

The flight crew of a CAP airplane, usually two men—a pilot and an observer—pictured here are Vole Smart avidly listening to Oliver Fullerton describing a flight. The work could be boring, as they spent hours flying over a convoy of ships.

This unidentified CAP member's half wing on his uniform indicates that he is an observer. The pilot and observer needed to work together as a team. Observers' tasks included maintaining navigational charts and looking for German submarines and downed military aircraft.

Planes that were patrolling the Gulf for submarines usually traveled in pairs. This was vital, as they were piloting single-engine airplanes designed for use over land, not over water. When one plane got in trouble, they radioed their sister ship for assistance. The sister ship's help could include circling a downed ship, dropping a life raft, and keeping in radio contact with the base.

These are some of the things that CAP members took on a flight. These included a life jacket, a dye pack, a camera, binoculars, navigational aids including a protractor and a clipboard to record data, and hunting knives. Pilots and navigators prayed that a long knife would enable them to fend off the hungry sharks and barracudas lurking in the Gulf of Mexico.

A CAP pilot holds headphones attached to radio equipment, which enabled him to remain in contact with the base. Crews were expected to report their location every 30 minutes. CAP pilots, who were technically civilians, had to pass a test from the Federal Communications Commission to obtain a license before operating radios.

Orren "Red" Walden holds a two-way radio. Walden had built radio stations for many companies in Dallas. The Civil Air Patrol actively solicited men with experience in both radio operation and repair. Building and maintaining a radio at the time required skill and ingenuity, especially at Base 10, where it was a hand-built operation. The base initially used a low-frequency radio but later upgraded to a CAA radio on 221 megahertz.

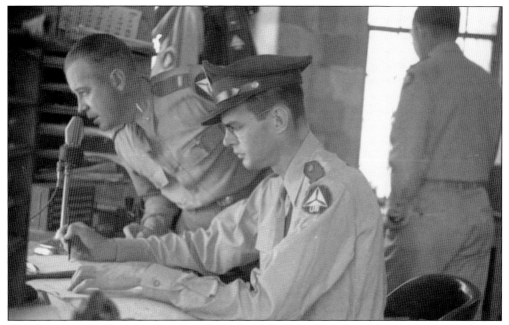

At left, a doctor believed to be Julian Fertitta speaks into the microphone as James Tomasson, a native of Arkansas, takes careful notes, probably recording the location of an airplane. Before the war, Tomasson had worked as a writer for the *Pine Bluff Daily Graphic* and the *Arkansas Democrat*. After the war, he returned to journalism, working as an Associated Press staff writer.

Observer Dean Hammond speaks on the microphone, probably letting the home base know their flight's location. After serving with the CAP, he was in the Army Air Corps. As a civilian, he was a pilot for private corporations until 1960, when he became a commercial pilot. He retired from Northwest Airlines in 1980. In 2006, he was awarded the Wright Brothers Master Pilot Award for over 50 years of flying without incident.

TSgt Mary B. Jackson and TSgt Thomas Vincent (or T.V.) "Pop" Connor work in the radio department of Base 10. Connor was born to Irish immigrant parents in 1874 in St. Louis. He was a mining engineer living and working in locations such as Nevada, Arizona, and Mexico. Pop Connor died in Dallas in 1961.

First Lt. Joe F. Long, a native of Wichita Falls, poses for the camera. Pilots were given a search area based on landmarks, sea buoys, radio beam waypoints, or latitude-longitude coordinates, and ordered to fly a back-and-forth grid or other search pattern.

Base 10 commander George E. Haddaway (left) and 1st Lt. Gilbert A. Mapes look at photographs; behind them is a plotting board, a blackboard that listed each pilot and their location.

From left to right, Donald Luce, Wesley W. Coffey, and an unidentified CAP member are seen here with the plotting board in the background. When the war began, Coffey wanted to serve, but he was too old to be drafted. He had extensive experience flying planes but had not been meticulous at licensure, so he was tested and made an observer. After the war, he moved to Dallas, where he briefly worked for James Marshall.

A pilot and observer carefully plan their flight path by utilizing a course plotter, a ruler with a compass marked for various map scales, and sectional and world aeronautical charts. A course would be plotted on a map, and estimates for winds, checkpoints, fuel burn, and more would be made.

Lt. Elbert C. Isom displays tools used for navigation, including an aeronautical chart, course plotter, and flight log. He was no ordinary CAP member, but the holder of a CAA ground instructor rating in navigation. Lieutenant Isom was a veteran of World War I and had 15 years of service in the US Navy and the Navy Reserve with a commissioned rank. He could speak, read, and write German and Spanish; speak and read French; and read many other foreign languages.

Rois Lamont Brockman (left) and Jack Neal are preparing for a flight in a Stinson Reliant. Brockman was an attorney in Dallas, Beaumont, and Woodville who was an integral part of the social life of each city. His wife, Willie J. Brockman, was the Beaumont city manager in the 1950s. At Base 10, he served as an administrative executive officer.

An unidentified CAP member inspects or maintains a Stinson Voyager 10A in a hangar. This type of airplane, a high-wing, single-engine aircraft, was often used by the CAP. It had a wingspan of 34 feet, a maximum weight of 1,965 pounds, and could achieve a maximum speed of 115 miles per hour.

Elmon F. Brockman, the older brother of Rois L. Brockman, was born in Sanger, Texas, in 1904 and grew up in San Angelo. At the time of his death in 1978, he was a pilot in his business, Brockman Aero Service.

This airplane, which was produced by Cessna as a DC6, features the nose art of a shark. Although nose art had appeared as early as World War I, the second world war was truly the golden age of the art form. George Haddaway referred to this airplane as a "circus wagon."

Many of the best pilots at Base 10, such as J.K. West, pictured here, were crop dusters. After the war, J.K. and his brother Earl wowed the crowds at air shows with their tied formation flying, with a cord tied to the wings of each plane as they went through a series of maneuvers without breaking it. West also seeded clouds to bring rain to drought-stricken rice fields.

This Stinson trimotor, possibly an SM-6000, could carry the pilot and 10 passengers. It was a powerful plane, boasting three 215-horsepower engines.

An unidentified CAP member hand props a Fairchild aircraft. Hand propping involved eye contact and shouting between the pilot and the person doing the propping. Note the CAP insignia, which came in three sizes: 20-inch for the fuselage, 30-inch for the wing, and 8-inch for automobiles. It featured a white triangle in a blue circle with a red, three-bladed propeller at the center.

Walter Strassmann Menge poses in front of a Stinson SM-6000. He was educated at Staunton Military School in Virginia and Cornell University in Ithaca, New York. After the war, he owned and operated a small airfield, Leon Valley Airport in San Antonio. Later, he flew for several airlines, including World Airways. In his retirement, he worked for an engine oil analysis firm.

Pictured is a Cessna C-34; it was a four-seater and was not particularly fast, but known for its aerodynamic efficiency. Like many other planes, it was built in Wichita, Kansas, known as the Air Capital of the World for its heritage of designing and constructing aircraft.

Thomas Vincent Connor Jr. and his father, also named T.V. Connor, who were known as "Doc" and "Pop," respectively, came from Dallas to Base 10. The younger Connor was a well-known Dallas dentist who showed films in his office to entertain and lessen the anxiety of his patients. The younger Connor died in 1960, and his father followed less than a year later in 1961.

These two photographs show Base 10 members learning how to use bombs by first using 100-pound M38A2 practice bombs. They were constructed of light sheet metal in the form of a cylinder, eight inches in diameter. The practice bomb was a powder blue color to differentiate it from an explosive-filled bomb. Importantly, it was filled with sand, not explosives, to simulate dealing with a real bomb. It was vital to learn how to maneuver an airplane with the added weight of a bomb attached.

This World War II poster urges CAP members to "Spot 'em, Plot 'em, and Pot 'em," meaning to locate a German submarine, plot their exact location utilizing dead reckoning or the location of a previously known object, such as the 18-mile lighthouse, and then "pot" or blow up the submarine with bombs.

D. Harold Byrd, a Texas Wing commander and founder of the Civil Air Patrol, proudly points to the patch showing a CAP airplane firing on a German submarine. In January 1943, Byrd reported that "the CAP planes are doing a heroic job." Previously, he had proclaimed to the *Dallas Morning News* on August 21, 1942, that "We feel that we have the most important waters adjacent to the United States to cover. We're getting plenty of action. We're doing our best. We're hunting the biggest game there is, and we're finding it."

As a Beaumont newspaper reporter quipped, "CAP men face three foes every time they fly—the axis, the elements and sharks." When the Civil Air Patrol began, it was only tasked with flying cover over ships and looking for survivors of submarine attacks or military airplane wreckage. Since they were flying so low and slow, they did spot items that faster military aircraft could not. When they made visual contact with a German submarine, they were to radio military authorities of the position and activity. The subs were taught to evade airplanes at a long distance by swiftly moving out of the area; however, if they were directly overhead, they were to rapidly sink below the surface, slow down, and let the airplanes burn precious fuel.

First Lt. Elbert C. Isom had a fascinating life. Born in 1896 in Michigan, he grew up in a home with live-in servants as the son of the president of Sinclair Refining. During World War I, he served as an officer in the Naval convoy service. After the war, he returned to Sinclair Oil and conducted an investigation into Europe and the Soviet Union. Then he studied engineering in Germany and returned to Sinclair in an executive and technical capacity. He was appointed commander of the Northeast Region of the Civil Air Patrol with the rank of colonel and later to the Civil Air Patrol national executive committee. In 1961, he became the vice president of the Barco Corporation, an oil and gas producing company. He was the holder of several patents pertaining to petroleum and navigation, and wrote numerous articles on petroleum engineering.

Pilot Buck Pratt, pictured here, and his observer had a close call when he was piloting his Stinson Voyager about 24 miles offshore. The motor suddenly started to make a grinding sound. He began to broadcast "99," or mayday, over the radio. But he changed his mind and continued to land on the beach. When Base 10 mechanics arrived, they found that the crankshaft had broken diagonally and was held together only by the main bearing.

First Lt. Edgar Kimball Jr. wears aviators, or pilot's glasses, which were developed in 1936 to protect pilot's eyes while flying. The aim of the glasses is to prevent as much light as possible from reaching the eye at any angle, so they feature dark reflective lenses with an area two or three times the size of the eyeball, and very thin frames.

Pilots relax and pose with an airplane. J.K. West, standing on the tire, at right, was a crop duster after the war and owned a surplus B-25 Mitchell medium bomber. He flew for what was then known as the Confederate Air Force, now known as the Commemorative Air Force, where he was known for his ability to fly anything.

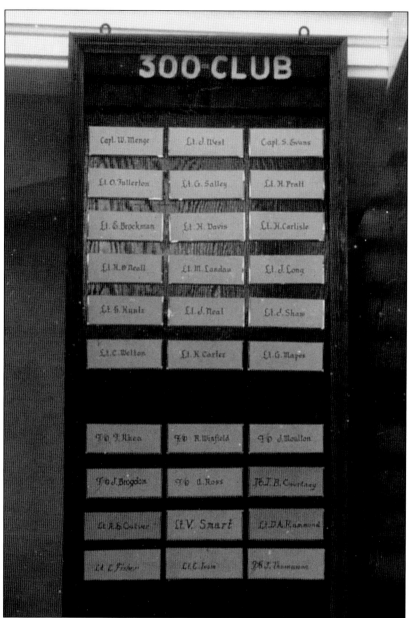

Base 10's 300 Club honored any pilot or observer who logged 300 hours official flight time over water on actual patrol, with their name posted on this sign and a check for $25. In 1949, forty-four men from Base 10 who flew 200 or more hours over water were honored with the Air Medal. They were R.V. Abshire, C.E. Barrett, J.C. Box, R.J. Bradley, Eldon F. Brockman, C.W. Brogdon, J.M. Brogdon, Hugh Carlisle, Kenneth J. Carter, Wesley Coffey, J.B. Courtney, Randall Edwin Culver, H.H. Davis, Charles Didio, Sumner C. Evans, Lawrence Fisher, John L. Fitch, Oliver Fullerton, Dean Hammond, E.C. Isom, William Jackson, G.S. Johnson, Edgar H. Kimball, Jr., E.J. Kuntz, J.F. Long, Donald F. Luce, Gilbert A. Mapes, James W. Marshall, Walter Menge, Joe Marshall, J.R. Moulton, J.C. Neal, R.M. Nicol, H.B. O'Neall, H.L. Pratt, F.A. Rhea, C.J. Roden, A.L. Ross, J.H. Shaw, Vole G. Smart, J.L. Thomasson, J.K. West, and Richard H. Winfield. Medals were posthumously awarded to Clarence P. Welton and Morris Landau.

The 18-mile lighthouse, also known as the Sabine Bank Lighthouse, was 18 miles south of Port Arthur. During the Civil Air Patrol days, pilots were delighted to spot it, as it indicated that they had only 18 more miles to the shore. It was constructed to warn ships of a sandbar, known as the Sabine Bank, where oceangoing vessels could get stuck. Construction of the lighthouse began in 1903, and after many adventures and expenditures of over $100,000, the tower was completed. The keeper and three assistants first beamed the light from its fixed, third-order Fresnel lens in early 1906. Given its remote location, keepers often served weeks or even months before being relieved. It was difficult to recruit and retain lighthouse keepers. In 1922, the problem was solved by activating an acetylene lantern in a Fresnel lens, ending the need for a keeper. The Sabine Bank Lighthouse was temporarily occupied during World War II as a coastal lookout station. The Fresnel lens was removed in 1971, and the tower was converted to solar power. In 2002, the top section of the lighthouse tower and the lantern room were dismantled and placed on display in Lions' Park in Sabine Pass, Texas. The original third-order Fresnel lens is exhibited at the Museum of the Gulf Coast in Port Arthur.

CAP members, mostly pilots and observers, pose in front of an airplane. This photograph may have been taken before a Sunday inspection. Although pilots and observers were all white men, they came from a variety of locations and occupations.

Thomas Vincent Connor Jr. was born at Gold Field, Nevada, in 1910. He lost his left leg in an airplane crash in 1944, in which the pilot was his CAP comrade Morris Landau. But in 1945, Connor saved himself and three others when a boat capsized, hauling the boat to an island to be rescued. His life was further complicated when he was in a car accident on January 23, 1950, in which his right leg was injured.

These CAP pilots in dress uniforms are, from left to right, John A. Toups, unidentified, Joseph F. Long, Vole G. Smart, Thomas V. Connor Jr., and Clarence P. Welton. Connor, a well-known Dallas dentist, makes a joke of phony teeth.

A CAP member is washing a Waco Cabin Class S. It was essential to wash an airplane after a mission, as flying at low altitudes over water resulted in a splatter of insects and a coating of corrosive saltwater spray. Hangars, as seen here, and water tanks at airports were often painted checkerboard red-and-white to help pilots identify the airport or confirm a checkpoint when heading toward another destination.

35

Wesley Coffey and his wife, Ruth Sanger Coffey, had a colorful life. Although he was born in Iowa and she in Philadelphia, Pennsylvania, they met in the Chicago area, where they worked as barnstormers during the 1920s. Barnstormers were aviators with tiny airplanes who would use barns as their venues for impromptu air shows, which featured aerobatics such as loops, rolls, spins, dives, and even wing walking, at crowd-thrilling, extremely low altitudes. Coffey, Ruth, and another man worked as a team. One would pilot the airplane, one would do wing walking acts, and the third would "pass the hat," asking people for donations. After one act, they only received 50¢. They decided to leave the business and eventually moved to Wichita Falls, Texas, where Wesley worked for an oil company. After the war, Ruth went back to school and became a professor at Lamar State College of Technology.

Ruth M. Coffey (right) and her nine-year-old sister Ann joined their mother and father, Ruth S. Coffey and Wesley W. Coffey, in Beaumont. Their parents were busy with work at Base 10, and so the girls spent much time alone. "Little Ruth," as she was known, was serious about her studies and graduated from high school at age 15. She was not just a scholar, but a red-blooded all-American girl who had her eye on the youngest pilot at the base, 18-year-old Donald F. Luce, pictured below. He was also from Wichita Falls, Texas. After the war, the family moved to Dallas, and Ruth began dating Luce. Although he had been discharged from the military in 1945, he returned to military aviation, and the Luces moved 25 times in 25 years and raised three children.

James W. Marshall, known as "Trappo," was an asset to Base 10 due to his lengthy experience with aviation and sound equipment. He was characterized by Comdr. George E. Haddaway as "a promotor and a wheeler-dealer, [an] excellent pilot with lots of crack up time." This implied that he was a good pilot who took too many chances. Born in Joplin, Missouri, in 1910, Marshall spent much of his life in Dallas and Beaumont. At the beginning of the war, he headed the Love Field Squadron at the Dallas Aviation School, which he brought to Base 10. He had also worked in sound for many movie theaters, including the Jefferson Theater in downtown Beaumont, where he worked with William R. Armstrong. Haddaway said Marshall was a master at radio work, who was dubbed "the wire twister." He collaborated with "Red" Walden and Armstrong in setting up the radio system at the base. In fact, Armstrong donated his ham radio set to Base 10 and then went to serve the military in another capacity.

Two

Spit and Polish

Stephen Sinkler and Pop Connor pose outside a building labeled "Smithsonian Institute," joking about the old and obsolete equipment that they used. Despite a lack of great material, Base 10 flourished thanks to the talent and ingenuity of the people who worked there.

One of the problems at Base 10 was that paychecks were small and, worse, late in arriving. How did CAP members survive? Some young members depended on modest handouts from family. Sometimes sympathetic members of the community, such as grocers, gave CAP employees a line of credit. In other cases, Comdr. George Haddaway co-signed loans at the bank. Some CAP members flourished because their expenses were so low. One group from Emporia, Kansas, comprised of two college professors, a radio shop owner, and a high school student, came to Base 10 for a summer, where they lived in the hayloft of a barn and took baths in a nearby canal. While some had vehicles and enjoyed evenings of fun entertainment, others never even went into Beaumont, which was only 10 miles away.

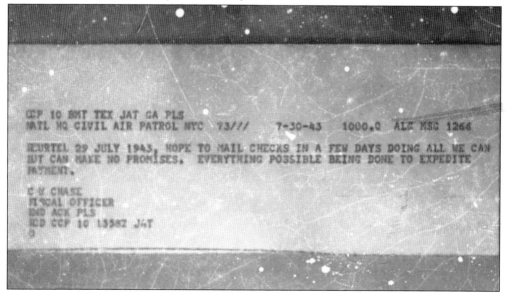

One of the things that distinguished the CAP from other military forces was that it was made up of civilians. About 25 percent of the pilots were Beaumont citizens, and like their counterparts in the Revolutionary War, they returned to their homes at the end of the day. There was even a book about the Civil Air Patrol named *Flying Minute Men*. As these photographs show, CAP members lived modestly at Base 10. George Haddaway recollected that when he brought his wife and infant child down from Dallas to live at Gallinipper Gables, it was a tourist court with lots of mosquitos. While many of Base 10's officers and their families lived on-site, in the barracks at Tyrrell Park or in trailers, enlisted personnel had to find housing at other places. Some lived in Hotel Beaumont, paying $1 a night. Others could not afford that and bunked in boardinghouses or at the YMCA.

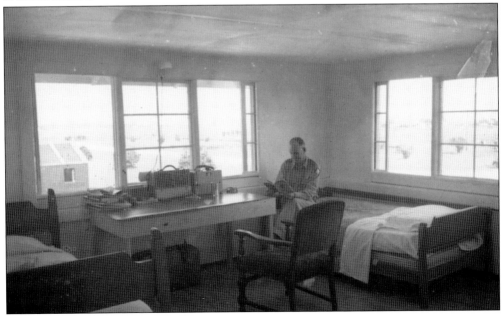

A tropical storm hit Beaumont on August 21, 1942, with 60 mile-per-hour winds and nine inches of rain between 7:00 a.m. and 6:00 p.m. Although damage in town was minor, with leveled garages, broken windows, and debris in the streets, the Civil Air Patrol faced the most harm. Three airplanes at Base 10, belonging to pilots from Texas, Kansas, and Nebraska, were severely damaged. This was a difficult loss, as Base 10, while not even two months old, had already lost five airplanes. Texas Wing commander D. Harold Byrd pointed out the need for additional hangar space to hold and protect 20 to 25 airplanes. He did not criticize the City of Beaumont, which he realized did not have the money, but instead appealed to oil and lumber companies to donate materials for the hangars, as the Civil Air Patrol was protecting their interests.

Charles Didio, with his back to the camera, a Navy veteran of World War I, worked to transform a group of middle-aged, overweight civilians into a professional corps of men who marched seamlessly on parade. George Haddaway characterized him as "absolutely a scholar in military drills, courtesy and discipline."

Sebastian Charles Didio, pictured at far right instructing the troops, was born in New York City in 1900 to Italian immigrant parents. After serving in the Navy, he moved to Texas, where he was an award-winning hairstylist who owned the Charles Beauty Salon in Beaumont.

Second Lt. Charles Didio, pictured here with a Piper aircraft, was the airdrome officer. He was responsible for maintaining the entire airfield area, including ensuring the safe takeoff and landing of aircraft, maintaining navigational aids, performing inspections, and communicating with air traffic control. His job entailed inspecting all airport areas, including hangars, runways, and fuel storage areas, for compliance with regulations.

George Haddaway examines troops who are standing at attention. Many of the CAP members were not young men but men ineligible for military service due to age or physical handicaps.

Base 10 commander George E. Haddaway, accompanied by 2nd Lt. Joe A. Marshall, examines the troops as Texas Wing commander D. Harold Byrd, at far left, looks on. In February 1943, Byrd and other Texas CAP executives came to Base 10 to examine members on their adherence to 30 different directives.

This image was captured during Sunday inspection, with the men marching in front of Koym-Taylor Hall. Many members of the community came to see the impressive ceremony, which was followed by a church service at the recreation hall.

Seen here from left to right are CAP national commander Maj. Earle L. Johnson, Maj. Harry K. Coffey, and Base 10 commander George Haddaway. Johnson, who was six feet, four and a half inches tall and weighed 240 pounds, was left guard for the Ohio State University football team in 1914–1916, where he was nicknamed "Tiny." While Johnson was on the football team, Ohio State won its first Big Ten championship.

From left to right are Charles Didio, Earle L. Johnson, and Randall Culver. When Undersecretary of War Robert J. Patterson expressed his interest in visiting CAP bases, The *Beaumont Journal* quoted Earle Johnson: "Well, if you visit any at all I would like for you to see the one at Beaumont, Texas. It is the outstanding one in the country."

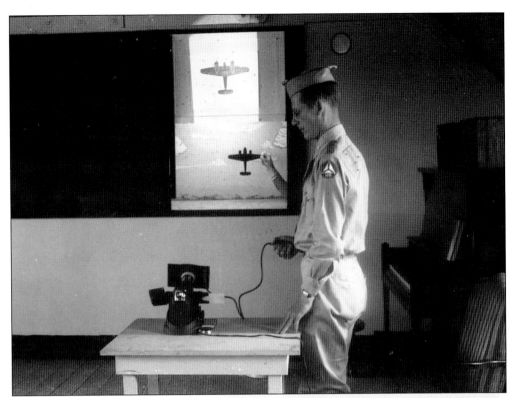

CAP members had to take 80 hours of basic instruction in topics such as meteorology, navigation, and general servicing of aircraft. Early in 1942, the classes were held in downtown locations, including Hotel Beaumont. Later, in the summer of 1942, Lamar College, today's Lamar University, announced that it was holding ground school training courses for the Civil Air Patrol. The classes, which were open to both men and women, were held Tuesday and Friday evenings from 7:30–10:30. Above, a CAP member shows slides demonstrating aircraft identification. Note that the slide is displayed by squeezing a bulb by hand. At right, Richard H. Winfield demonstrates how to identify ships while flying overhead.

Above, Larry Jean Fisher instructs his students. Teaching others was not a new task for Fisher, as he had worked in entertainment since he was a teenager, including directing vaudeville acts, emceeing children's events at the Jefferson Theatre, and producing a historical play. Below, an unidentified CAP member demonstrates vacuum trajectory, or utilizing ballistics to calculate velocity. This would be an aid in determining how to drop bombs that would hit a German submarine.

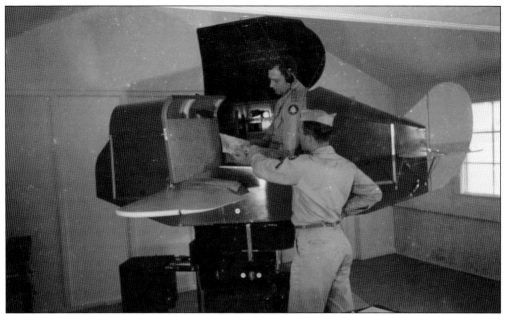

CAP members, including Alpha Boger, in the photograph below, learn about aviation on a Link Trainer, a flight simulator developed by Ed Link in the 1920s as a safe way to teach pilot trainees how to fly by instruments. More than half a million pilots were schooled on Link Trainers. George Haddaway realized the importance of a Link Trainer to Base 10 and heard from his good friend Maj. Sidney Price, commander of the Lake Charles Army Air Force Base, that one was available. This was great news, as Link Trainers were invaluable and much in demand. Unfortunately for Base 10, the trainer was routed to the 25th Antisubmarine Base in Galveston. Haddaway found out that the trainer was going to be shipped out of the Houston Freight Yard, so he sent some of his men to Houston, where they lied and said that it was supposed to go to Base 10, and hauled it to Beaumont. Although Haddaway admitted he "got into Hell" over his action, it provided badly needed training to Base 10 pilots.

Joe P. Klein displays the red epaulets, or shoulder straps, which distinguished a CAP uniform from Army apparel. The straps had a buttonhole, which was attached to a plastic button with the insignia of the patrol. In Beaumont, in the fall of 1942, two Red Cross sewing units answered an urgent call for 150 epaulets in a very short time. According to an interview with Del Gallier, "No one wanted to wear the epaulets. They were pretty conspicuous. They stood out like a sore thumb. But we had to wear them."

An aviation superstar, Lt. Comdr. Philip Van Horn Weems came to inspect Base 10 on September 30, 1942. Weems was not only a US Navy officer, but also a genius who invented navigational instruments and techniques that revolutionized aerial navigation, and authored books on navigation. One of Weems's most noteworthy inventions was the Weems Second-Setting Watch, which had a rotating 60-second bezel, which enabled the military to synchronize their watches and coordinate troop movements down to the second. As he was known as the expert on navigation, Charles Lindbergh consulted Weems when he returned from his historic solo flight across the Atlantic. The famed aviator realized that he would need more navigation expertise to tackle even more dangerous flights. Weems taught him to use celestial navigation to determine position. Pictured from left to right are unidentified; Lt. Comdr. P.V.H. Weems; Maj. D. Harold Byrd, commander of the Texas Wing of the CAP; and Comdr. Phil Hamps, port director of Galveston.

When Weems came to Base 10, the conversation centered around two topics: aviation and football. Commander Weems, a noted proponent of the Civil Air Patrol, proclaimed that the CAP halted submarine activities and saved a half-billion dollars' worth of cargo a month and the lives of innumerable Merchant Marines. He believed it could change the course of the war. Weems took an active role in getting vital supplies to Europe, as he served as a convoy commander, for which he was honored with a Bronze Star. But as this photograph shows, the visit was not all serious. Commander Weems, who attended the US Naval Academy in Annapolis, was an all-American center for Navy the same year that Commander Byrd's first cousin, noted explorer Richard E. Byrd, was a quarterback. Their coach that year was none other than Comdr. Phil Hamps of Galveston.

One of Base 10's bragging points when Lieutenant Weems visited in September 1942 was the acquisition of the famed aircraft *Spirit of Africa and Borneo*. This was not just any garden variety airplane, but a Sikorsky S-39, an amphibious craft that could float in the water as well as fly in the air. It had been owned by celebrities and Kansas natives Martin and Osa Johnson, pictured here in 1935. The Johnsons were well-known adventurers who studied the animals and native peoples of Africa and Asia and shared their remarkable journeys through films, books, and magazine articles. Osa, a celebrated beauty, made several best-dressed lists and had her own line of clothing. The aircraft boasted giraffe patterns accurately depicting a particular species of giraffe. (Both, courtesy of the Martin and Osa Johnson Safari Museum, Chanute, Kansas.)

CAP members, including Jimmie "Trappo Jr." Marshall in aviator glasses and Claud W. Moody at right, pose in front of the *Spirit of Africa and Borneo*. In 1932, the Johnsons learned to fly and commissioned two planes, an S-38 and an S-39, from the Sikorsky factory in Connecticut. Igor I. Sikorsky's career spanned 60 years, and he is credited with many achievements, including the development of the helicopter. Osa piloted the *Spirit of Africa*, photographing wildlife from the air. The Johnsons were the first photographers to fly over Mount Kenya and to film Mount Kilimanjaro. They flew their Sikorsky S-39 over 60,000 miles. The *Spirit of Africa* was especially remarkable, as only 21 aircraft of this type were ever built. There are three remaining, one of which was painstakingly restored by Dick Jackson of Rochester, New Hampshire. He found the hull of NC-50V in the Alaskan bush and spent 40,000 hours over four decades restoring it to its former glory. Today, it is in the Florida aviation museum Fantasy of Flight.

Three

Tragic Losses and New Solutions

A Coast Guard ship tows the famed *Spirit of Africa* after a failed attempt to rescue CAP pilot and observer James Charles Taylor and Alfred H. Koym on November 11, 1942. Unfortunately, a line broke, and the aircraft sank within 15 minutes into the Gulf of Mexico.

Pictured from left to right are Capt. George Haddaway, commander of Base 10; Lt. James W. Marshall; and Lt. Robert F. Neel. Marshall piloted the Sikorsky S-39 from Tucson, Arizona, to Base 10. The newly obtained *Spirit of Africa* was designed to be used as a rescue craft. It was much larger than other aircraft at Base 10, with a gross weight of 4,000 pounds and a wingspan of 52 feet. It was much more powerful as well, boasting a 300-horsepower Pratt & Whitney R-985 engine. When the magnificent craft sank into the Gulf of Mexico, Texas Wing commander D. Harold Byrd, who owned it, filed an insurance claim of $7,923.53

Two CAP members—Lt. James Charles Taylor of Baton Rouge, Louisiana, pictured below, and Lt. Alfred Herman Koym of Rosenberg, Texas, at right—died on a routine patrol flight on Veterans Day, November 11, 1942. It was Taylor's first week at the base and his first day as a pilot at Base 10. The plane was inspected and deemed fit for offshore patrol. It is speculated that Koym and Taylor's plane had a broken crankshaft in its Ranger engine. They were 30 miles from land over the Gulf of Mexico. Taylor maintained radio contact with the base and initially reported that he had landed safely on the water. But the men were in great peril, as they had no lifeboat; it was a cold November day, and Koym could not swim. While their life preservers kept their heads above the water, over time, temperature and fatigue took their toll.

There was a tremendous effort to save Taylor and Koym. The men on their sister ship, Walter Menge and Richard H. Winfield, immediately radioed the base for assistance. They remained with the downed plane, constantly circling it. Within 72 minutes of the accident, directed by the remarkably accurate information supplied by the sister ship, help had arrived. When word reached Base 10 of the downed aircraft, all the men wanted to go and rescue them. Experienced pilot and observer Robert "Wimpy" Neel, pictured at left, and J.K. West, above, were dispatched in the Sikorsky S-39. West climbed on top of the large plane to try and locate the men. Lieutenant Neel called for West but heard no response, so he turned off the engine.

In the heavy seas, the engine was sprayed with water and would not start again. Their ship was then in grave danger when a pontoon was knocked off the wing, and the wing ended up in the water. Neel and West were rescued by the Navy, which had been dispatched to the scene. They also recovered the bodies of Taylor and Koym, who died of exposure. An attempt was made to save the *Spirit of Africa*, but it had to be let go. Pictured here, CAP members hold flowers, which they dropped in the Gulf of Mexico at the site where their fellow CAP flyers A.H. Koym and James C. Taylor lost their lives on a mission. Walter Menge (left) and Richard H. Winfield (right) were on the downed plane's sister ship. Taylor was obsessed with motors, as he had a flying service with his brother Thomas H. Taylor, who served in the Civil Air Patrol in Omaha, Nebraska. He also operated Taylor Auto Repair Shop in Baton Rouge.

The loss of four men was a bitter blow to Base 10. As with Taylor and Koym, impressive ceremonies were held in honor of Dean and Ward. Full military funerals were held on November 18, 1942, with CAP members as pallbearers and CAP airplanes flying overhead. Public memorial services were held at Base 10 with a full-dress review and retreat. Joe Z. Tower, minister of the First Methodist Church, gave the address. Lena Milam, director of Beaumont music education, performed a violin rendition of "Ave Maria." Pictured is a model of Lt. John Henry Dean's plane constructed by 12-year-old Paul Enloe; it was covered with red carnations and posed for a westward flight.

The memory of Taylor and Koym was honored at a memorial service on November 15, 1942, at a ceremony where the recreation hall was renamed in their honor. Standing at attention above from left to right are Capt. Del Gallier, Capt. Robert Wallace, First Lt. Randall Culver, 2nd Lt. Joe A. Marshall, and Base 10 commander George E. Haddaway. Flight officer Wesley W. Coffey, standing on top of the stairs, acknowledges and returns their salute and unveils the plaque. After the full-dress review and retreat, Rev. George F. Cameron, rector of St. Mark's Episcopal Church, officiated at a service at an altar decorated with red, white, and blue flowers and American flags. Cameron's talk, "The Unnumbered Hosts," discussed those who lose their lives in military service. The service also included vocal selections and organ and piano numbers.

Two CAP members, Lt. John Henry Dean Jr. and Lt. Robert D. Ward (right), were killed, and a third lieutenant, Del Gallier (below), sustained a fractured ankle on November 16, 1942, only five days after Taylor and Koym died. Dean was piloting his Rearwin aircraft on a routine flight to pick up an airplane in Fort Worth. Visibility was bad, so he dropped down to verify where they were by reading the name of the town on the water tower. Dean then made a short climbing turn to get back on course. Unfortunately, the airplane was overloaded with three men and much luggage, and went into a spin, heading straight down at full power. The airplane hit trees, which took off the wings and tail section. It crashed at a lumberyard in downtown Mexia, Texas. The townspeople came immediately to help, but Dean and Ward could not be saved. Gallier spent a week in the hospital recovering from his injury.

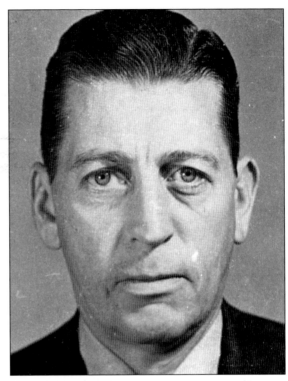

John Henry Dean Jr. is pictured at right with his airplane, a Rearwin. Also pictured are Les Stringer (left) and Walter S. Menge. Dean was a prominent Texas rancher, with extensive land holdings around Fort Worth and Lamesa. He was a descendant of a well-known cattle-raising family. He was born in Dallas in 1906 and educated at the Terrill School and the Culver Military Academy. When he was a teen, his father gave him the responsibility for purchasing range bulls for the ranch, and he became the manager of the ranch in 1928. He was superintendent of the National Hereford Show at the Texas State Fair for two years, as well as serving as vice president of the Texas Hereford Association. R. Dean Ward was born in Fayetteville, Arkansas, in 1942 and educated at the Kentucky Military Institute. As a young man, he operated an air school in Muskogee, Oklahoma, and accumulated 400 hours of flying time. He came to Dallas in 1932 and was employed by the McAlester Fuel Company until he joined CAP.

Cpl. John Meggs was an expert on the use and care of parachutes. A December 13, 1942, *Beaumont Enterprise* article discussed his work. He was quoted as saying, "Jumping is really very fun. It's something like riding a roller coaster or sliding down a snowy hill." Meggs was an Army veteran of seven years, whose training included attending the Army Parachute School. He later performed over 400 jumps in an aerial circus and was a stunt man in the MGM film *Parachute Nurse*.

Corporal Meggs discusses the importance of meticulously packing parachutes, which were 24 feet in diameter and composed of 65 yards of silk. A very long table was laid out with the material and lines to get them in perfect condition before jumps.

Corporal Meggs instructed CAP members not to jump backward out of an airplane but to instead jump feet first. CAP members had to be above 800 feet in altitude before they could jump out of an airplane. They also learned how to control a parachute in the wind, as seen here.

On November 12, 1942, just a day after Koym and Taylor lost their lives in the Gulf of Mexico, Comdr. George E. Haddaway was determined to get a lifeboat, even if he had to steal one. He and Randall Culver, Base 10 operations officer, set off for the 25th Antisubmarine Wing at Galveston. They had previously sent a CAP member to the base for supplies, who returned with the layout of the hangar and the location of the lifeboats. Haddaway knew that the guards left at lunch, so they arrived then. Although one guard remained, Haddaway distracted him while Culver loaded a lifeboat into their airplane, and they took off for Base 10. Everyone on base anxiously awaited their arrival, and a huge cheer arose when they returned with the precious cargo. The photograph below shows a lifeboat attached to the bottom of an airplane, which could be dropped to a sister ship in the water.

In 1942, four officers from Base 10 pose in front of the famed *Spirit of Africa* before a flight. From left to right are Comdr. George E. Haddaway, 1st Lt. Oliver P. Fullerton, 1st Lt. Francis R. Parkman, and Andrew Jackson McCauley. They are wearing life preservers with fluorescent dye markers tied on. If they were in the water, they could open the packets, and a fluorescent green dye would spread over the surface, increasing their visibility to an airplane or ship.

While some CAP members may have been cavalier about wearing a life vest, even throwing them in the backseat of the plane, after the deaths of Taylor and Koym, they realized their importance. Here, George Salley is ready for a flight with two life vests, a Mae West, and a kapok jacket securely tied around his torso. After the war, Salley returned to his home state of Alabama, where he unsuccessfully ran for governor.

Here, a CAP member wears the pin denoting membership in the Duck Club, for CAP members who had been rescued from an airplane that developed engine problems and landed in the water.

Pictured at Base 10 are, from left to right, pilot Charlie Kehoe, T.V. Connor, and Andrew Jackson McCauley with their deflated lifeboat. They were near the 18-mile lighthouse when they suffered engine failure, but were picked up uninjured in two or three hours. The Civil Air Patrol had long been aware of the need for lifeboats and were ingenious in obtaining one. In a July 1942 Beaumont newspaper article, E.B. Germany requested that citizens, especially campers and fishermen, donate an easily inflatable rubber mattress to Base 10. Although they asked for donations, they also rigged one themselves by utilizing two rubber tires held together by canvas.

Pictured below on June 7, 1943, Capt. Sumner Clark Evans of Dallas, an Army Air Corps instructor during World War I, had resumed flying on July 2, 1942. Evans claimed that he had flown more than one thousand hours of antisubmarine duty. His observer, flight officer R.H. Winfield Jr. of Arkansas, is shown at left checking flight papers before takeoff. Their experience proved that with the right equipment, CAP members could survive an engine failure and a crash into the water. They were wearing life preservers and did attempt to get their life raft free, but their plane sank so quickly that they could not get it loose, so two life rafts were dropped to them from other CAP planes. They drifted seven long hours, from 6:00 p.m. Sunday to 1:30 a.m. Monday, in the darkness, 35 miles from Galveston. The Coast Guard was able to locate them by their whistling "Into the Deep." They were taken to Galveston, where they were picked up by Comdr. George E. Haddaway and returned to Base 10 unharmed.

Four
Partners for Victory

The engineering crew demonstrates the talent and ingenuity of CAP members who, as Larry Fisher, put it "kept the crates flying." While many would not have been accepted into the regular military because of age or physical handicap, they were vital to the success of Base 10.

Capt. Walter Strassman Menge was born in 1912 at Dolgeville, New York. His father was an immigrant from Germany whose family owned a shoe company. Before the war, he worked for Edgar Tobin Aerial Surveys at San Antonio. He died in California on July 6, 1988, while learning to hang glide.

Operations officer 1st Lt. Randall E. Culver (right) and his assistant Walter Menge are pictured here. One of Culver's accomplishments as operations officer was recognizing the need to secure airplanes at Base 10 before a tropical storm, which hit Beaumont in 1942. He buried cement blocks with steel eyes into the ground at various points to be used as tie-downs. Other operations officers included Gus Whiteman, Robert Wallace, and Leslie Stringer.

William F. Van Cleave, born in 1867, was perfect as the assistant engineering officer, as he was working on planes before many at the base were born. He was raised on a farm in Ellis County, Texas, and claimed in a 1928 article in the *Dallas Morning News* that he "got out of the cradle with a wrench in his hand" and grew up in his father's blacksmith and machine shop. As a young man, he fought in the Spanish-American War and repaired American aircraft during World War I. Van Cleave, who moved to Dallas in 1903, was watching an electric fan in 1908 when he conceived the idea to build an airplane. He was successful and began a career designing and building aircraft, including a 21-foot midget plane in 1928.

One of the challenges for Base 10 was that they could only use 50 percent of the hangar, as the other half was utilized for storing civilian pilots' planes. Texas Wing commander D. Harold Byrd had big plans for the hangar, desiring to add a 100-foot extension to provide additional storage.

A member of CAP inspects an aircraft engine with a rocker and plugs in the foreground. Mechanics were highly prized at Base 10, as many were drafted into the other services.

This is an airplane oil sump pan with excess sludge. Sludge was caused by not changing the oil frequently enough. Later, detergents were added to the oil. The cylinder is a bearing that would have been at one end of the assembly. It is a unique setup.

William VanCleave, who had decades of experience building airplanes, stands at far left with a bandsaw, which was used to cut wood into a contoured shape for an airplane body. The saw was powered by a small electrical motor.

CAP members are performing maintenance work on a Fairchild 24C-8C. The two poles at left in the background alternated red and white and were antenna masts for a non-directional beacon, a wire stretched between the two poles. This was a navigational aid to guide aircraft to the airport and to shoot instrument approaches in poor weather.

Jake Baker Jr. was one of the African Americans who were vital to the success of Base 10. World War II was a time of great change for African Americans, with higher wages and new job opportunities, but many difficulties remained, with a race riot taking place in Beaumont in the summer of 1943.

Edgar E. Duncan, at right in the first row, wore a loaded Colt .45 when he went to pick up African American cooks, pictured below in the canteen with a propeller clock, and janitorial staff from their homes during the Beaumont Race Riot in 1943. His wife, Mabel, pictured above at far left in the second row, supervised the group. The riot erupted on June 15, 1943, and ended on June 17. It was precipitated by a huge influx of workers into Beaumont, with the population exploding from 59,000–80,000 from 1940 to 1943. Thousands of people were now competing for housing and jobs in defense work, especially the shipyards. An executive order forbade racial discrimination, so blacks and whites were competing for jobs. The riot illustrated the worst of American life on the home front. In the end, one black man and one white man were dead, 50 people were injured, and 200 were arrested.

The kitchen for Base 10 shows an open door letting in badly needed air. Cooking, in the days before air conditioning and microwave ovens, was a hot and labor-intensive process. Fires were a constant danger, hence the fire extinguisher. The canteen provided three meals a day. It operated like a regular café, with customers ordering what they liked and paying at the cash register. The canteen was one of four buildings, including a recreation hall and two barracks that were moved to the base from Tyrrell Park, where they had been used by the Civilian Conservation Corps in the 1930s.

Mabel Duncan, operating the cash register, gives Larry Fisher 35¢ in change as her husband, Edgar Duncan, looks on approvingly. The couple were married on Christmas Day in Floydada, Texas, in 1937. After the war, they operated a Piggly-Wiggly grocery store in Hamlin, Texas.

Officers enjoying breakfast in the canteen are, from left to right, unidentified, George Haddaway, Gus Whiteman, and an unidentified CAP member. Note the jukebox behind Haddaway and the pinball machine. Like many other things, pinball machines were in short supply during World War II.

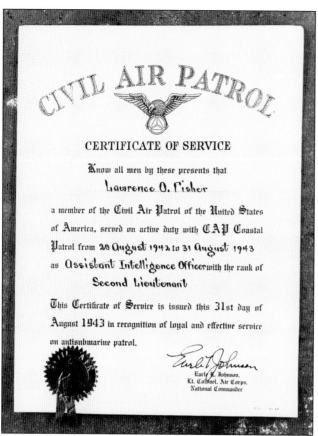

Larry Jene Fisher's certificate of service as an assistant intelligence officer is seen here. Intelligence officers coordinated activity between the Civil Air Patrol, initially part of the Office of Civilian Defense and branches of the military, including the Army and Navy. Intelligence officers at Base 10 included E.B. Germany, Burton Holton, and Glen S. Ramsey.

Bernard Burton Holton was an aviation enthusiast from a young age. In the early 1930s, he was a pilot and president of the Wing Over Club, a group of pilots who worked to improve aviation facilities, create flying events, and improve public perception of aviation. He was also the nephew of Gladys Bingham, namesake of Gladys City, an oil boomtown that sprung up during the Spindletop boom of the early 20th century.

Jimmie Marshall (left) and Robert Wallace are pictured here. Marshall was known as "Trappo Junior," as he was the son of James W. Marshall and was only 14 when he served as an aide to Comdr. George Haddaway, who remembered him as someone "who flew like a duck, swam and served in every capacity from yard bird to morale builder." Marshall had grown up in the cockpit of his father's airplanes and had soloed in an airplane before he had his driver's license.

Helen Culver works on the financial records of Base 10. She was born in Virginia in 1905 and married Randall Culver; they lived in Marcus Hook, Pennsylvania. Although she did not work outside the home in 1940, she remained in the workforce after the war and was a diligent employee.

Base 10 guard duties were to secure the base and protect property and the lives of those who worked there. Guards were posted at the front gate, the terminal building, the armament shack where the bombs were stored, and in the hangars. James Parsley remembered that "all enlisted buck private through master sergeant were required to stand guard duty." One night when Parsley was on duty, he saw a car approaching and, following the rules he had been taught, attempted to stop the vehicle coming in without being recognized, but the car did not stop. Parsley, following orders, "took both rear tires off [with a 12 gauge shotgun] and some other damage as well." Parsley continues, "Well, out of the car roared our base commander and I was dressed up one side and down the other and I don't recall any repetition. I was standing at the most severe attention repeating Yes sir, Sorry sir, ad infinitum." Haddaway ordered him to report to his office at 8:00 a.m. He was in total fear but was relieved when the base commander said, "Parsley, you were right, I was wrong, and we both have something to remember, just don't do it again."

At Base 10, women played an important role, but they were not pilots or observers. Four of the women who worked at the base are pictured below; from left to right are Alpha Boger, Helen Culver, Mary Ellen Box, and Mary Roden. Except for Boger, all their husbands were officers at Base 10. Their work included accounting, operating the plotting board, and bookkeeping. African American women provided a vital service by cooking in the canteen and cleaning throughout the base. The Civil Air Patrol would not let women fly in the coastal patrol, but many did fly courier services ferrying men and badly needed equipment to military installations and armament factories. In fact, there were some CAP squadrons composed entirely of women.

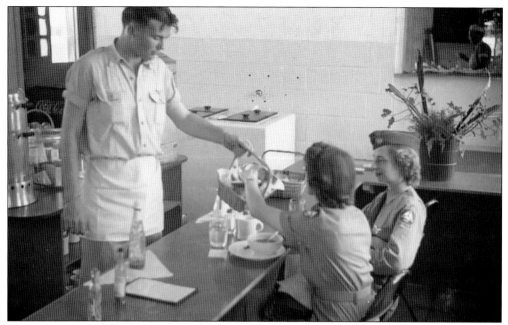

A woman CAP pilot relaxes with Helen Culver as Culver accepts a roll from a male CAP pilot. This pilot may have delivered important papers to Base 10. World War II brought dramatic changes to the role of women in Beaumont and throughout the country. Women were now entering the workplace and earning their own money.

World War II gave new opportunities to women. Now women, as well as men, were proud to wear the uniform of the Civil Air Patrol. Although women were not allowed to fly antisubmarine missions, many did learn important skills, which some utilized as WASPS.

Clarence P. Welton sits by a typewriter. Office work during World War II was complicated by a shortage of typewriters. Demand was higher than ever, with the mountains of paperwork created by the war and a need for quick and easy-to-read communication. But many typewriter companies had been asked to retool their factories. Royal, for example, was making airplane parts. So, like many other commodities, including tires, shoes, and sugar, typewriters were rationed.

John Toups is seen here with a teletype machine, which was used for the transmission of priority messages. Like many people in Southeast Texas, Toups had French Louisiana roots, as his father had come to Texas from Lafourche Parish, Louisiana.

TSgt John A. Toups speaks to an aircraft crew and records data about their location. Toups, born in Beaumont in 1920, was one of the younger people at Base 10. He had earlier worked as an office boy for a rice broker.

Clarence P. Welton types a report while smoking a cigarette. During World War II, smoking was an integral part of life, as people were mostly unaware of the health risks. In 1943, American companies manufactured 290 billion cigarettes. Soldiers were actually encouraged to smoke, as it relieved boredom and improved morale.

Five

A Nation at War

Base 10 was an integral part of the community. Many locals served at the base in many different capacities. Groups in the area pitched in to provide for the needs of the patrol, and the CAP assisted with community events that raised money for CAP or sold war bonds. Pictured are an unidentified pilot and Wesley Coffey (right) with Red Cross workers at a war bond sale on December 30, 1942, across the street from what is now the Tyrrell Historical Library.

Lamar Aviation School was part of a government-sponsored program that began in 1938 to teach young people how to fly to aid in military preparedness. What was then Lamar College worked with the Aviation Committee of the Beaumont Chamber of Commerce to lobby Congress for a program at Lamar. Their work was successful, and in 1939, the Civilian Pilot Training Program began with 72 classroom hours and 35 hours of flying instruction.

The *Spirit of Dallas* was a trim, $5,000 Stinson Voyager 10A purchased in a week by over 40 Dallas civic organizations. Citizens were also encouraged to send money; as the *Dallas Morning News* touted, a $5 contribution for a $5,000 airplane could save countless lives and $50,000 in goods being shipped. Moreover, they could take pride in a plane emblazoned with the name "Dallas" sinking an enemy ship.

The citizens of Southeast Texas worked hard to meet the needs of the Civil Air Patrol. One problem was providing transportation from Hotel Beaumont, where many of the men stayed, to the CAP base 10 miles away. The American Women's Voluntary Service (AWVS) solved the problem. They not only raised money to purchase a 1941 Plymouth station wagon, pictured here, but also provided drivers as part of the Motor Corps. The AWVS had a grueling 19-hour schedule, beginning at five in the morning and continuing until midnight. The drivers had to do much more than keep the vehicle out of the ditch. They had training in map reading, first aid and advanced aid, auto mechanics, and blackout driving. The Motor Corps adopted military protocol taught by Charles Didio, wore uniforms, and carried a knapsack containing first aid supplies, chocolate, cigarettes, and cosmetics.

Jimmie Marshall, a 14-year-old messenger, is pictured at left outside the office at Base 10; note the message on the wall behind him, "This Space for AWVS Wagon." The photograph below shows the AWVS wagon waiting to transport CAP members downtown. Their work at providing transportation was impressive, driving 21,000 miles and taking 8,500 CAP members to the base. The AWVS was the largest American women's service organization. The local chapter boasted 2,000 members and provided a wealth of services to the community, including a day-care center for the children of women working in war industries. Helen Ivers, unit chair of AWVS, said the organization offered women the opportunity to "stop being the little woman . . . and the chance to be trained, dependable, and capable."

Navy Day, October 27, 1942, featured the largest parade held in Beaumont. It was a four-mile parade with 1,500 participants and 60,000 spectators. The parade featured nine school bands and a wide array of military and civic organizations, including the Boy Scouts, the Beaumont Music Commission, Altrusa Club, the Daughters of the American Revolution, and the Young Men's Business and Professional League. At the beginning of the event, CAP airplanes swooped low over the city, with loud bursts from simulated "bombs," high-powered fireworks set off from the top of the San Jacinto Building. These photographs show members of the Civil Air Patrol waiting for the parade to begin, with 1st Lt. Gilbert Mapes and 1st Lt. Joe Klein (with his back to the camera below) exchanging a salute.

African Americans are pictured above as spectators of the Navy Day parade, and their houses are seen below. Blacks numbered 20,000, or one-third of the population of Beaumont, in 1941. According to Beaumont Chamber of Commerce records, their houses were "bare, leaky, and bleak shelters." The publication stated that many African American "citizens 'made-do' without sewer connections and only half the water supply required for domestic needs." In general, the streets where they lived were rutted and dusty in dry weather and quagmires in the rain. Worse, the community suffered an appalling infant mortality rate of 68.1 per 1,000 live births compared to whites, who suffered 13.4 deaths per 1,000.

Base 10 commander George E. Haddaway, flanked by Gus Whiteman (left) and Robert Wallace, leads CAP members in the parade. Nelson's Market, at 806 Park Street, which specialized in fish, is pictured at back left near the American flag. Note the enthusiastic youngsters peering from the window of the car.

Charles Didio, who taught CAP members marching and military protocol, leads the way. Note Beaumont Typewriter and Supply Company at 675 Orleans Street, which was providing a vital wartime service, as it not only sold typewriters but also repaired them.

From left to right, Gus Whiteman, Robert Wallace, and George Haddaway are ready for the parade. The winged badges on their hats indicate that they are officers. The trio illustrate the diverse backgrounds of CAP men; before the war, Whiteman operated a lumberyard in Alto, Wallace was a San Antonio attorney, and Haddaway published an aviation magazine in Dallas.

The chaplain, with a cross insignia on his hat, holds a sign reading, "Enlist the Seabees." The Seabees, or Construction Battalions, were formed in January 1942 by Rear Adm. Ben Moreel, as international law made it illegal for civilian workers to resist an attack, and as guerrillas, they could be summarily executed. The Seabees were members of the armed forces capable of any type of construction, anywhere needed, under any conditions or circumstances.

CAP members march on Pearl Street near what is today the Beaumont Public Library. A billboard touts Regal Beer, which originated in New Orleans's French Quarter in 1890. The American Brewing Company, which used the popular jingle, "Red beans and rice and Regal on ice," closed in 1962. The tower at back is the imposing 13-floor Art Deco Jefferson County Courthouse.

Base 10 commander George Haddaway, flanked by Capt. Gus Whiteman (left) and Capt. Robert Wallace, marches past what was then the Tyrrell Public Library. The stone building at left, a marvel of Richardsonian Romanesque Revival architecture, was constructed in 1903 as the First Baptist Church. It was repurposed in 1926 as Beaumont's public library. After the construction of a new library, it was converted to the Tyrrell Historical Library in 1974.

CAP members parade in front of the Perlstein Building on Pearl Street, which housed the S.H. Kress five and dime store on the bottom floor and professional offices on the second floor. Although the structure was demolished, a granite pillar still stands on the grounds of what is today the Art Museum of Southeast Texas. Note the huge crowd of spectators, estimated at 60,000, for Navy Day.

CAP members stand in front of what was then city hall and turn to stare at a 1941 Buick Century sedan. It got its name from the fact that it was guaranteed to be able to reach 100 miles per hour. They were the fastest Buicks of their era and were nicknamed "the banker's hot rod."

A war bond sales event was held on December 30, 1942, on the grounds of Beaumont City Hall, which sold $73,950 of war bonds in only 12 hours. A committee, chiefly composed of CAP members from the city, planned to wow the locals with aviation. The highlight of the show was the Sinclair Flight Trainer, furnished by the Sinclair Oil Company, which was an abbreviated airplane that was harnessed to the ground and powered by an electric motor. Here, a CAP member is trying out the simulator at Base 10 as other members peer out the windows. The tower of the building has radio antennae so the base could communicate with airplanes during their missions. The loudspeakers attached to the tower enable anyone at the radio controls to send messages to everyone on base, while the nearby cooling tower allowed chilled water to pass through a pipe, lowering the temperature of the building.

The main attraction was the Sinclair Flight Trainer; at full speed, the propeller blasted a 75-mile-per-hour wind over all control surfaces, giving the trainer an immediate response to the controls, as in a real airplane. It simulated any maneuver of an aircraft—climbing, leveling off, banking and turning, looping, and descending.

The flight simulator was shown in downtown Beaumont on the lawn of what was then city hall and is today the Julie Rogers Theatre. Visitors to the event viewed aviation equipment, which included demolition bombs, an airplane fuselage, and floatation gear, including a rubber boat like one used by CAP members in the event of water landings.

The crowd was fascinated by the flight trainer, which members of the public could fly. Note the loudspeakers attached to the automobile at right, a 1941 Packard sedan. The building at left is the old YMCA, erected in 1903 to provide a wholesome environment for young men during the first Spindletop oil boom. After a new Y was built in the 1920s, it sat vacant for many years.

A man films the Sinclair Flight Trainer in front of what was then city hall, which was constructed in 1927 in the wake of the riches of the second Spindletop. The hall boasted not only city offices, but also a 2,500-seat auditorium that hosted many entertainment greats, including John Philip Sousa, Paul Whiteman ("the King of Jazz"), Irish tenor John McCormack, and pianist Ignace Paderewski.

The most money spent on war bonds was by Elmon W. Doty, a special representative and assistant agency manager of the South Coast Life Insurance Company, pictured in the flight simulator, who purchased a $10,000 bond. Also pictured is Capt. H.S. McIntosh, group commander and general chairman of the drive. The local squadron of the Civil Air Patrol was composed of Beaumont citizens who acted as part-time auxiliary members of the CAP as opposed to active-duty members stationed at the airport.

The proceedings were broadcast on Beaumont radio station KRIC, and over $50,000 in war bonds were sold the first day of the event. The "V-5" painted on the airplane refers to a five-cylinder piston engine with the cylinders sharing a common crankshaft arranged in a V formation. The stone building with the Gothic tower was the Tyrrell Public Library, today's Tyrrell Historical Library.

The honor roll for the City of Beaumont commemorates men who were killed in the war. Many of them have an "MM" after their name, indicating that they died in the Merchant Marine. The AWVS compiled a list of the men who had lost their lives when the organization was notified of the loss by the deceased's love ones, and placed the names on a sign in front of city hall for the duration of the war.

Base commander George E. Haddaway enjoyed his time in the flight trainer. Haddaway remarked on Civil Air Patrol flights over the city that simulated bombing, "When you hear an airplane over Beaumont, that is your cue, Mr. and Mrs. Beaumonter to 'fly' to city hall and buy a bond at the CAP bond booth." The most recognizable emblem on the simulator is the famous Sinclair dinosaur logo "Dino." It was first used in Sinclair's marketing in 1930 to sell lubricants refined from crude oil. Overall, the event sold war bonds worth $73,950.

Randall Culver is being handed a five-gallon can of aviation fuel by an unidentified CAP member. Interestingly, Culver worked as a foreman at a Sinclair refinery in Pennsylvania before the war. The Sinclair Oil Company recognized the likelihood of impending war, and took strategic steps to allow the company to grow during World War II. One of its most important moves was anticipating the need for 100-octane gasoline and experimenting with alkylation and polymerization processes beginning in 1937. By the beginning of the war, Sinclair was able to quickly expand its production of aviation gasoline.

On February 13–14, 1943, a Japanese midget submarine captured at Pearl Harbor in 1941 came to Beaumont on a war bond sales tour. The Civil Air Patrol was involved with the event, with Base 10 commander George E. Haddaway on the planning committee and the CAP marching unit in the parade, which brought the 81-foot sub on a tractor-trailer to downtown Beaumont. It had previously been shown in 162 cities throughout the nation. Here, the sub is pictured with an enthusiastic crowd at Orleans Street and Laurel Avenue. The building at right, then a furniture company, had previously been Nathan's Department Store. It was built in 1910 for Jacob J. Nathan, who was born in 1874 in Indiana to Russian-born parents. Like many merchants of that era, he was Jewish. Nathan was a retail innovator who aimed to always offer the lowest price on his goods. In a time before television, the Internet, or even radio, he got his message across by plastering signs reading, "Nathans Sells It for Less" on every post, fence, or tree in a 50-mile radius of Beaumont.

The submarine's arrival in town was a major event, as it was routed through the city as part of a big parade. Participants included the South Park High School Band and their Greenie Cadets, French High School Band and their Victory Corps, Lamar College Girl Corps, American Legion and the American Legion Auxiliary, Business and Professional Women's Club, Catholic Daughters of America, Beaumont Music Commission, Altrusa Club, Council of Jewish Women, Beaumont Trades and Labor Assembly, and the Boy Scouts. After its capture, the sub was stripped of all equipment, so it was taken to Mare Island Naval Shipyard, 25 miles northeast of San Francisco in Vallejo, California, where it was restored to fighting trim by the addition of duplicate batteries, periscope, dials, and gauges.

The crew of the submarine was ordered to enter Pearl Harbor, attack two American warships with torpedoes, and then scuttle this sub with explosives. Plans went awry, however, as they were unable to enter the harbor and ran aground. After a long ordeal, one of the men drowned, and the other was captured. The sub was transported to the submarine base at Pearl Harbor, where it was examined, providing a wealth of technical data and documents. In 1947, it was put on display at the naval station at Key West, Florida. In 1964, it was moved to an outdoor exhibit at the Key West Lighthouse and Military Museum. In 1990, the museum began focusing on the lighthouse and divesting itself of military collections. The following year, the sub was moved to Fredericksburg, Texas, where it is displayed at the National Museum of Pacific War.

An enthusiastic crowd views the 81-foot sub being transported through the city. The event was a huge success, raising $400,000 in war bonds. Admission was a 25¢ stamp for children and $1 for adults. Two iconic stores of the city are pictured here, including the national chain Sears and the dry goods retailer the Fair Store. It was operated by two Jewish merchants of Prussian descent, first cousins Benjamin Greenberg and Benjamin Dorfman. The store remained in operation for generations, adding a branch at the Parkdale Mall in the 1970s. The downtown store remained in operation until the early 1980s.

A photographer stands on a car to take photographs of the Japanese midget submarine. He is in front of Hotel Beaumont, at the corner of Orleans and Fannin Streets. The hotel was of vital importance to Base 10, as many men from the base stayed there or met their rides on the AWVS van there. The 11-story hotel, which opened in 1922, was the first million-dollar hotel in Beaumont. It was an important gathering place from the 1920s through the early 1960s. One of its most luxurious features was the Rose Room, which hosted proms, parties, weddings, and professional meetings. Over the years, the hotel was home to many businesses, including the KFDM radio station, a cigar store, and the Kitten coffee shop. Business declined over the years, and the hotel changed hands 10 times between 1966 and 1974, when it was converted into a retirement home. In 1995, a campaign began to restore the hotel and provide quality affordable housing for 135 senior citizens. Unfortunately, the hotel closed in 2011, but Hotel Beaumont is prime real estate and, at the present, is open for re-development.

The Japanese midget submarine is seen here parked on Fannin Street in front of the Jefferson Theatre during a scene of frenetic activity, with three organizations—the Red Cross, the AWVS, and the Young Men's Business League—frantically selling tickets at booths outside the Jefferson. It was estimated that 1,000 people an hour passed over the walkway attached to the outside of the sub, peering inside, as windows had been cut into the steel hull. The Jefferson was where residents loved to while the hours away. The million-dollar theater opened in 1927 to a crowd of 5,000 who wanted to see the movie palace. It had features including six types of marble, chandeliers 18 feet in diameter, statues of Greek gods, and a water fountain as expensive as a new Ford automobile, but what was truly amazing was that it boasted a $56,000 Carrier air conditioning system. The first air-conditioned building in town was a huge attraction, especially in Beaumont's hot, humid climate. Over the years, the Jefferson has been the place to see vaudeville acts and both silent movies and "talkies," as well as a wide array of musical performances.

Lt. (jg) Frances Rich (second from right) came to Beaumont on May 27, 1943, to stage a drive for women to join the WAVES, the women's branch of the US Naval Reserve during World War II. In this photograph, she is pictured with, from left to right, Comdr. George E. Haddaway, who was on the committee that planned her visit to Beaumont; an unidentified naval officer, an unidentified WAVE, and Charles Didio. Rich was the daughter of Maj. Charles Rich and silent-screen star Irene Rich. After graduating from Smith College in 1931, she played secondary roles in six Hollywood films and on Broadway. But it was sculpting—not acting—that was her true passion in life. She pursued her love of sculpting at Cranbrook Academy of Art, where she was a resident student from 1937 to 1940. One of her most famous works is an 11-foot Tennessee marble Art Deco statue honoring nurses who died during military service, located at Arlington National Cemetery. When World War II came, she initially served as a draftsman at an aircraft factory but soon enlisted in the WAVES and served as a special assistant to the director.

Roscoe Ates, a popular character actor in vaudeville, movies, and television, is pictured second from left above in Base 10's canteen. He is with, from left to right, Miles Smart, John "Hoss" Moulton, and an unidentified man. Below, Ates is pictured with Charles Didio. Ates was best known as the character Soapy Jones in Westerns. He was born in 1895 in a small Mississippi town where he was embarrassed by a severe speech impediment. As a youth, he played the violin accompanying silent movies at a theater in Chickasha, Oklahoma. He soon moved on to performing as a concert violinist but found more opportunities as a comedian in vaudeville, where he utilized his stutter for humorous effect. During World War II, he served in the Air Force fighter squad program at Ellington Field in Houston.

Six

THE RENAISSANCE MAN OF EAST TEXAS

Larry J. Fisher was vital to Base 10. He preserved the history of the base as the archivist/historian, photographing everyone at the base, even a pet skunk, at work and play. He also gave presentations to community organizations, taught courses to other pilots, and flew more than 300 hours over water. After Base 10 ceased antisubmarine operations, he remained with the Civil Air Patrol, fighting forest fires and producing films.

Larry Jene Fisher was born Lawrence Orsino Fisher on June 18, 1902, on a ranch near Wichita Falls. His father had worked as a jeweler and watchmaker for decades but was forced into bankruptcy, and after a brief interlude living near his in-laws in Dallas, he departed for California, where he died in 1927. His mother struggled to support herself and her son by working as a boomtown hotel keeper, a milliner, and a private nurse. When Lawrence was only 15 years old, he embarked on a career as an organist at silent movie theaters, performing in many different states. He was the consummate showman, adopting the stage name Larry Jean Fisher, organizing vaudeville performances, and when not in the Lone Star State, performing as the "Texas Organist," as depicted in the caricature below. Around 1946, he changed the spelling of his middle name to Jene.

Larry Jean Fisher was chosen as the organist for the Jefferson Theater in downtown Beaumont in 1931. The Jefferson, pictured here in 1947 for a premiere of *It's a Wonderful Life*, was a glamorous theater created in the wake of the second Spindletop oil boom. It boasted a $30,000 Robert Morton Wonder Organ with 778 pipes, custom designed and voiced for the theater. It was shipped from Van Nuys, California, via the Panama Canal. It could replicate a wide variety of sounds, from a cello to a car horn. The huge organ was installed near the orchestra pit on a disappearing platform, which was operated by a 10-horsepower engine that boasted enough wiring to build a 75-mile fence. Fisher not only performed daily at the theater, but also managed the Organ Club, which offered entertainment for children on Saturdays. (Courtesy of the Rolfe and Gary Christopher Negative Collection, Special Collections, Lamar University.)

Fisher developed a passion for flying while working at the Jefferson Theater. He obtained a pilot's license in 1931 and quickly became a leader in local aviation. He was a member and president of the Wing Over Club, which brought new adventure to his life. Fisher performed in many air shows; at one, he stood on a wing playing the accordion while Lois Neel, pictured here, piloted the plane.

Larry Fisher left the Jefferson at the end of 1934 to pursue a career in aviation. He offered many services, including aerial photography, flying lessons, charter trips, and air ambulance, but aviation did not pay the bills. Fisher returned to music, giving accordion lessons at four different studios in Beaumont, Port Arthur, Lake Charles, and Jennings. He and his most accomplished students, the Accordionaires, gave performances at a wide array of venues.

While Larry Fisher was flying at low altitudes over Texas, he noticed the Big Thicket and how the lumber industry was changing the environment. While experts disagree over exactly what the boundaries of the Big Thicket are, most agree that it is a densely wooded region north of Beaumont. Larry Fisher, pictured with his camera on horseback, was also a photographer and was intrigued when he heard that the Thicket region boasted a wealth of orchids.

Fisher was so fascinated with the Big Thicket that he moved to the heart of the region, Saratoga. Here, he established a photography studio where he took portraits of area residents. Fisher stayed up late developing film, while his neighbors rose early and blared country and western music from their radios. Fisher hated country music and exacted revenge by blasting classical music at a loud volume.

Larry Fisher is pictured at center directing *Keyser Burnout*, a play he wrote and produced in 1940–1941 as an attempt to document Big Thicket history. In the play, Texas men evade serving in the Confederate army by hiding in the woods before a Confederate officer tries to literally smoke them out. The play had features authentic to the Big Thicket, including cast members who portrayed one of their ancestors, and artifacts including a 150-year-old spinning wheel.

Here, Fisher shows color slides of the Big Thicket to a standing-room-only crowd in East Texas. He was an officer of the East Texas Big Thicket Association, which wanted to create a Big Thicket national park. Fisher recognized the importance of the area with its diversity of plant and animal life and traditional folkways. He documented the region with thousands of photographs used in newspaper articles, exhibits, and a book.

At Base 10, Fisher, seen here with his back to the camera, was the archivist/historian. His photographs and words prove that he realized the important role everyone at the base played, from CAP elite to the mechanics, guards, plotting board operators, clerks, and cooks. The images of the Civil Air Patrol taken by Fisher at Base 10 are preserved and made accessible at Lamar University's Special Collections.

This photograph spoofs a rumor that Larry Fisher (far left) was a German spy operating a high-powered radio transmitter in the depths of the Big Thicket, busily sending messages to the Nazis. The gossip persisted, and Fisher discovered that it originated with a woman who worked at the White House, a Beaumont department store.

Fisher (kneeling at center) was frustrated that some people in the community labeled the Civil Air Patrol as "loafers" when he knew the difficulty of being a pilot flying for hours over convoys of ships carrying cargo desperately needed by the Allies, constantly watching for German subs. He was delighted to share photographs and stories that set the record straight in Robert E. Neprud's *Flying Minute Men: The Story of the Civil Air Patrol*.

After men from Base 10 were no longer chasing German subs, Fisher remained with the Civil Air Patrol, where he worked to establish the Texas Forest Service's forest fire patrol and headed its audio-visual department. He rose to the rank of major. Always on the cutting edge of technology, he worked as a filmmaker and in television, as shown in this photograph of puppets from a children's program. Ironically, the man who hated country music died on July 6, 1953, in Nashville, Tennessee.

Seven
No Dull Boys at Beaumont

Base 10 was a rowdy place. People who spent their days endangering their lives by flying old single-engine aircraft at low altitudes over the Gulf of Mexico needed to blow off steam. The men pictured here are an example of some of the antics at the base.

Drinking alcohol was not allowed on the base, so these CAP members enjoyed a brew at a tavern after marching four miles in the Navy Day Parade on October 27, 1942. Perhaps Base 10 commander George Haddaway, at center, treated his men to the first round.

CAP members enjoyed dropping coins into the slot machines seen here. In fact, the "one-armed bandits" rescued the canteen, which had been operating at a loss. Here, officers including, beginning second from left, George Haddaway, Earle Kunz, and Gus Whiteman enjoy a meal.

CAP members had to find out how a lifeboat and life preserver would really work in the water, so they went skinny dipping at the swimming pool at the YMCA on Calder. Several of the men from Base 10 lived at the Y because it was cheaper than living at Hotel Beaumont. The YMCA, which features Spanish-style architecture, was built in the 1920s in the wake of the second Spindletop, an oil boom that spawned a bonanza of beautiful buildings. Today, the old Y serves as housing for senior citizens.

Romance was alive and well at Base 10, as seen in these photographs. Joe Richard Wier of Opelousas, Louisiana, is pictured above with an unidentified sweetheart in downtown Beaumont on December 30, 1942. At left, an unidentified pilot poses with his love interest. As CAP member Joe N. Summers recollected decades later, "On our off-duty time, we chased the girls. When you are young and your hormones are working right, what else is there?" While many relationships in Beaumont may have been flings, some became serious. A wedding of a CAP member was held on September 1, 1942, that reflected CAP's status, as it was described as having a "semi-military atmosphere." It was held at Hotel Beaumont and featured a reception hosted by wing commander D. Harold Byrd with a cake made at Base 10's own canteen. The best man was Comdr. George E. Haddaway, and his secretary Mavis Gallier was maid of honor. While that particular marriage later floundered, other Base 10 romances stood the test of time.

Pets added levity to the stressful work at Base 10. Animals shown in Larry Fisher's pictorial record of the base include wiener dogs and a white cat. Pictured here is a pet skunk apparently owned by pilot George Salley (above). Below, the skunk is being fed by Evelyn Tatum in front of a Stinson 10A Voyager. But it was a black cat known as Three Point who lived on in stories of the base. Before the cat was commissioned as a flight officer in a solemn ceremony, Three Point was content to hang out with the guards and mechanics. But once he received a promotion and a bar was attached to his collar, he began to lounge and stand in formation with his officer friends.

These photographs illustrate relaxed times at Base 10. Below, men are enjoying a story told by "Hoss" Moulton. Note the map in the background showing the Gulf coast of Texas and Louisiana, where the men made their flights. While some of the younger men enjoyed playing softball, the most popular pastimes at the base were playing cards and dominoes. Pictured above are "Red" Walden, at center facing the camera, and Joe Marshall at far right. Willis F. Rose Jr. recollected that he liked to go to Port Arthur's Pleasure Pier to hear the famous bandleader and trumpeter Harry James perform. As he put it, "I liked to dance, and I never found any shortage of girls."

Eight
Moving on to New Work

In late summer 1943, the scourge of German submarines was over, and the last flight left the base on August 31. This photograph shows the flag being lowered on that day, marking the end of an era. The Civil Air Patrol had played a key role in ending the menace from submarines. By September 1942, shipping losses were reduced to just one ship, with only seven more lost during the rest of the war. Bases from Bar Harbor, Maine, to Brownsville, Texas, had flown 86,685 missions, with a total of 244,600 hours of flight.

Base 10 commander George E. Haddaway, who was awarded the War Department Exceptional Civilian Service Medal, poses with a rolled-up map. All the CAP members were presented with a photograph album entitled "C.C.P. Base 10 Pass in Review," with pictures and captions created by Larry Fisher. Haddaway also wrote a letter, which read in part, "No matter where you go in this world you can hold your head high with pride in the knowledge that you served your country in the field of combat during the great crisis of 1942–1943. You may be decorated for your active duty, for your valor, and for your services, but no matter the honor or the decoration that might come from higher authority there will be no reward that can compare with the inward satisfaction from having served on CAPCP Base 10, and the knowledge within you that you are a part of a great victory."

After Base 10 closed, many CAP members, especially younger members, continued to serve their country in the armed forces. Some were in roles directly related to their CAP work, and others utilized skills they had honed at Base 10 as military and civilian pilots. For example, Charles Didio worked in a similar position as a commander of the traffic management operating system at a naval base in New Orleans. A Coast Guard captain, out of gratitude for his good work keeping the subs at bay, gave Kenneth Carter and Don Luce (pictured here) the opportunity to study celestial navigation and other marine subjects at an air transport command school in New Orleans. As Carter recollected, "In six weeks we were commissioned second lieutenants and assigned to sea-going tugboats." Another CAP member who had a noteworthy career was Edgar H. Kimball Jr., a fighter pilot in the Pacific who participated in every major amphibious operation in that theater of the war. Serving at Base 10 gave some CAP members skills they utilized not only as military pilots, but as pilots of commercial airliners for many decades after the war.

Maxine Johnston is pictured with the Larry Jene Fisher Collection at the library at what is today Lamar University in Beaumont around 1970. She served in many different capacities at Lamar, from reference librarian to library director (1980–1988). Among her many contributions to the library was the creation of a department of special collections emphasizing the Big Thicket.

BIBLIOGRAPHY

Blazich, Frank A. Jr. *"An Honorable Place in American Air Power": Civil Air Patrol Coastal Patrol Operations, 1942–1943*. Maxwell Air Force Base, AL: Air University Press, 2020.

———. " 'Definitely Damaged or Destroyed': Reexamining Civil Air Patrol's Wartime Claims." *Air Power History* 66, No. 1 (Spring 2019): 19–30.

Civil Air Patrol documents including "Air Medal Study," "Certificate of Belligerency Coastal Patrol Personnel," and "Qualifications of CAP Coastal Patrol Officers and Flight Personnel." Louisa S. Morse Center, Washington, DC.

genealogybank.com.

Geroux, William. *The Mathews Men: Seven Brothers and the War Against Hitler's U-Boats*. New York, NY: Penguin Random House, 2016.

Keefer, Louis E. *From Maine to Mexico: With America's Private Pilots in the Fight against Nazi U-Boats*. Reston, VA: COTU Publishing, 1997.

Larry Jene Fisher Collection. MS 3, Mary and John Gray Library, Lamar University, Beaumont, TX.

Mellor, William B. Jr. *Sank Same*. New York, NY: Howell, Soskin, 1944.

———. *Flying Minute Men: The Story of the Civil Air Patrol*. New York, NY: Duell, Sloan and Pearce, 1948.

newsbank.com.

newspapers.com.

Offley, Ed. *The Burning Shore: How Hitler's U-Boats Brought World War II to America*. New York, NY: Basic Books, 2014.

Price, Gladys. Interview by Jan Street, November 6, 1990. Lamar University History Department, Oral History Collection, MS 63. Mary and John Gray Library, Lamar University, Beaumont, TX.

Walker, John H. Papers, AC-241, Tyrrell Historical Library, Beaumont, Texas.

Wiggins, Melanie. *Torpedoes in the Gulf: Galveston and the U-Boats, 1942–1943*. College Station, TX: Texas A&M University Press, 1995.

About the Organization

The University Archives and Special Collections is an integral part of the Mary and John Gray Library at Lamar University, where the history of both Lamar and Southeast Texas is preserved and made accessible.

Special Collections features original materials, including sound recordings, films, photographs, letters, diaries, and rare books that document the unique story of the region.

A particularly intriguing part of Southeast Texas is the Big Thicket. Although experts disagree on the boundaries of the region, most agree that it is primarily a wooded area north of Beaumont. Its special qualities include rich biodiversity of plant and animal life, such as four of the five carnivorous plants of North America, and its romantic heritage of traditional ways of life. Lamar's Special Collections contain dozens of collections documenting the Big Thicket's environment and culture, but the largest is the Big Thicket Association papers compiled by Maxine Johnston, which reveal a wealth of information on the Big Thicket, including the work to create the first national preserve and the efforts to expand and improve the Big Thicket National Preserve.

A major watershed in the history of the Big Thicket and all Southeast Texas was World War II. This is documented in Special Collections with papers that illustrate both the home front and Texans' experiences in the war while serving in many different capacities around the globe.

An amazing treasure in Special Collections is the Dishman/Justice Cookbook Collection, which includes 1,639 volumes and whose importance lies not in its size but its rarity and age, with some volumes dating as early as 1500. The not only document cuisine, but also provide an intimate glimpse of life in previous times through home remedies, etiquette guidelines, and instructions for the management of servants.

Lamar University's Special Collections document the dramatic, dynamic, and diverse history of Southeast Texas through over 200 notable collections.

PO Box 10021, Beaumont, Texas, 77710
409-880-7787 or 409-880-8660
www.lamar.edu/library/services/university-archive/index.html

Discover Thousands of Local History Books
Featuring Millions of Vintage Images

Arcadia Publishing, the leading local history publisher in the United States, is committed to making history accessible and meaningful through publishing books that celebrate and preserve the heritage of America's people and places.

Find more books like this at
www.arcadiapublishing.com

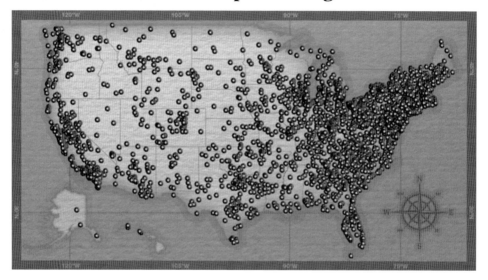

Search for your hometown history, your old stomping grounds, and even your favorite sports team.

Consistent with our mission to preserve history on a local level, this book was printed in South Carolina on American-made paper and manufactured entirely in the United States. Products carrying the accredited Forest Stewardship Council (FSC) label are printed on 100 percent FSC-certified paper.